TOMORROW AI

THE FUTURE, REIMAGINED

RAMESH PINGILI

To my family, whose unwavering support and boundless patience fueled this journey into the fascinating and sometimes bewildering world of artificial intelligence. This book is a testament to our shared explorations and late-night discussions that illuminated my path, especially when the algorithms seemed to have their own minds. It is also dedicated to the countless researchers, innovators, and dreamers who have devoted their lives to pushing the boundaries of what is possible, shaping a future where human ingenuity and artificial intelligence intertwine to create a technologically advanced and profoundly humane world. Their unwavering commitment inspires us to strive for a future where AI is a powerful force for good, solving some of humanity's most pressing challenges while safeguarding our shared values. To those who grapple with the ethical implications of AI and work tirelessly to ensure responsible development — your insights are invaluable, and this work contributes to that crucial ongoing conversation.

Contents

Foreword

The rapid evolution of Artificial Intelligence has captured our era's imagination and fundamentally transformed how we live, work, and think. As a technological phenomenon, AI has transitioned from being a subject of speculative fiction to becoming an integral part of our everyday reality. Yet, as we stand on the precipice of this revolution, questions about its potential, risks, and the future it promises remain ever more pressing.

"Tomorrow AI" is a comp" telling explore "action of these questions—a thoughtful journey into the heart of AI's transformative power and its possibilities for humanity. The author unpacks the layers of AI's impact on industries, society, and individual lives through a meticulous blend of research, expert analysis, and visionary insights.

This book explores AI's technological marvel and philosophical and ethical dimensions. It challenges us to consider the human values that must guide its development and adoption, ensuring that this extraordinary tool enhances, not diminishes, our shared humanity.

As someone deeply invested in the intersection of technology and society, I find this book timely and essential. It bridges the technical depth required to understand AI's mechanisms with universal implications. Whether you're a seasoned professional navigating the complexities of AI in your field or simply a curious reader pondering what tomorrow holds, this book will spark meaningful reflections and discussions.

"Tomorrow AI" is more than a book—it's an invitation to think critically, dream boldly, and prepare purposefully for the future that AI will help shape. I invite you to turn these pages with an open mind and a keen sense of curiosity as we embark on this exploration together.

Ramesh Pingili

IEEE Senior Member, Fellow of IETE, Full Member of Sigma Xi

Technologist and Author Specializing in AI and Workflow Optimization

Preface

Artificial intelligence is no longer a futuristic fantasy; it's a tangible reality interwoven into the fabric of our daily lives. From the algorithms that curate our newsfeeds to the AI-powered medical diagnoses that save lives, this transformative technology is reshaping our world at an unprecedented pace. Yet, amidst the excitement and rapid advancements, a crucial need remains for a clear, accessible, and nuanced understanding of AI's implications. 'Tomorrow AI' is written for everyone – the tech enthusiast, the curious general reader, the concerned citizen, and the future leader. This book does not shy away from the complexities of AI, but it avoids overly technical jargon, employing relatable examples and compelling narratives to illuminate even the most intricate concepts. The aim is not only to inform but also to inspire critical thinking and informed discussion, to cultivate a sense of responsible engagement with this rapidly evolving field. The journey through 'Tomorrow AI' is intended to be more than just an acquisition of knowledge; it is an invitation to participate in shaping the future of artificial intelligence to ensure that this powerful technology serves humanity's best interests, protecting our values and upholding a future that is both innovative and equitable. Join me as we explore the possibilities and challenges of this fascinating world together.

Acknowledgements

Writing *Tomorrow AI* has been an incredible journey, and I am deeply grateful to everyone who contributed to its realization. First and foremost, I want to express my heartfelt appreciation to the brilliant minds in the AI community whose pioneering work and insights have shaped the depth and relevance of this book. Their relentless pursuit of innovation continues to inspire and push the boundaries of what artificial intelligence can achieve.

I am also immensely grateful to those who supported me throughout this journey—my family, friends, and mentors—whose encouragement and belief in this project provided the motivation to see it through. Their unwavering support has been instrumental in bringing this book to life.

Lastly, I extend my gratitude to the countless researchers, innovators, and visionaries whose dedication to artificial intelligence continues to shape our future. Their work not only informs this book but also lights the way for the next generation of AI advancements.

Prologue

The dawn of Artificial Intelligence marks a turning point in human history—a moment where imagination and innovation converge to redefine our existence. What was once the realm of science fiction has now seeped into every corner of our lives, from the devices we hold to the systems that govern industries, cities, and nations. AI is no longer a distant possibility; it is our present reality and, more importantly, our future.

As we stand on the cusp of this profound transformation, we face both exhilarating opportunities and daunting challenges. Will AI empower humanity to solve the most pressing global issues, from climate change to healthcare? Or will it unravel the social fabric through unchecked biases, economic disruptions, and ethical dilemmas? These are not hypothetical questions but the decisions shaping our collective tomorrow.

"Tomorrow AI" is my attempt "to illuminate" the path ahead. This book is not just about what AI is or how it works; it is about what AI means for all of us. It is a journey through its vast potential, the risks it carries, and the choices we must make as creators, consumers, and stewards of this transformative technology.

In these pages, we will explore AI's far-reaching impact across industries, its ethical and philosophical dimensions, and its role in shaping a future that is not only technologically advanced but also fundamentally human. Together, we will ask the following questions: What kind of future are we building? Who will benefit from it? And critically, how do we ensure that AI becomes a force for good?

We do not passively await the future—we actively create it. As you embark on this exploration, I invite you to imagine a world where innovation is guided by wisdom, progress is tempered by responsibility, and technology catalyzes a brighter, more equitable tomorrow.

Welcome to Tomorrow AI. Let's envision the future, let's.

CHAPTER ONE

The Genesis of AI

The Genesis of AI

Historical Overview of AI

The journey of Artificial Intelligence began in the mid-20th century with the aspiration to create machines that could mimic human intelligence. The spark for this vision can be traced back to Alan Turing's groundbreaking 1950 paper, "Computing Machinery and Intelligence." In this seminal work, Turing introduced the concept of the "Imitation Game," which later became known as the Turing Test—a method to evaluate a machine's ability to exhibit intelligent behavior indistinguishable from a human's. Turing's audacious question, "Can machines think?" laid the philosophical and technical foundation for AI research.

The 1956 Dartmouth Conference, organized by John McCarthy, Marvin Minsky, Nathaniel Rochester, and Claude Shannon, is widely regarded as the birthplace of AI as a formal field of study. At this historic event, McCarthy coined the term "Artificial Intelligence," envisioning a future where machines could perform tasks such as reasoning, learning, and self-improvement. Early pioneers like Herbert Simon and Allen Newell developed programs such as the Logic Theorist, which could prove mathematical theorems, and the General Problem Solver, showcasing the early promise of AI systems.

While the early years of AI were optimistic, the journey was far from linear. Funding and interest fluctuated as researchers grappled with the complexity of replicating human cognition, leading to periods of stagnation often referred to as AI winters.

Key Breakthroughs and Milestones

1. **1950s-1970s: Foundations and First Programs**

- AI's Birthplace: During this period, AI was primarily conceptual, focusing on problem-solving and symbolic reasoning.
- Early programs, such as ELIZA, developed by Joseph Weizenbaum in 1966, demonstrated rudimentary natural language processing. ELIZA mimicked a psychotherapist's conversational style but exposed AI's limitations in understanding context.

Despite these early achievements, the lack of computational power and data restricted AI's growth.

2. 1980s: Expert Systems

- The rise of expert systems, such as MYCIN, marked a shift towards simulating specialized human expertise. MYCIN could diagnose bacterial infections and recommend treatments, showcasing AI's potential in medical diagnostics.
- This era witnessed a surge in corporate investment, driven by promises of practical applications in industries ranging from healthcare to manufacturing. However, high development costs and limited adaptability eventually tempered enthusiasm.

3. 1990s: The AI Revival

- The 1997 victory of IBM's Deep Blue over world chess champion Garry Kasparov was a watershed moment. This achievement demonstrated AI's capacity for strategic thinking, albeit within narrowly defined parameters.
- Simultaneously, advancements in statistical learning and data availability laid the groundwork for modern machine learning.

4. 2000s-Present: Machine Learning and Neural Networks

- The emergence of deep learning, spearheaded by projects like Google Brain, revolutionized AI. Neural networks grew more profound and complex, enabling breakthroughs in tasks like image recognition and speech synthesis.
- AI applications expanded dramatically, powering innovations such as virtual assistants (e.g., Siri, Alexa), autonomous vehicles, and real-time language translation.
- The confluence of big data, enhanced processing power, and improved algorithms propelled AI into mainstream adoption.

The Foundational Technologies Behind AI
Machine Learning (ML)
ML, a subset of AI, enables machines to learn patterns from data without explicit programming. Techniques include:

- Supervised Learning: Models learn from labeled data, such as predicting housing prices from historical sales data.
- Unsupervised Learning: Models uncover hidden patterns in unlabeled data, such as clustering customer preferences.
- Reinforcement Learning: Systems learn through trial and error, optimizing behavior based on rewards (e.g., AlphaGo mastering the game of Go).

Neural Networks
Inspired by the structure of the human brain, neural networks consist of interconnected layers of nodes (neurons) that process information. Key advancements include:

- Convolutional Neural Networks (CNNs) for image processing.
- Recurrent Neural Networks (RNNs) for sequential data like speech and time series.
- Transformer Models: Recent innovations like GPT and BERT have redefined natural language processing.

Data Science

- The abundance of big data and advancements in data processing fuels AI's modern success.
- Data Science integrates algorithms, statistical methods, and computational techniques to extract actionable insights.
- Cloud computing platforms and distributed systems, such as Hadoop and Spark, have enabled the processing vast datasets at unprecedented speeds.

The Foundational Technologies Behind AI

Machine Learning (ML)

ML, a subset of AI, enables machines to learn patterns from data without explicit programming. Techniques include:

- Supervised Learning: Models learn from labeled data, such as predicting housing prices from historical sales data.
- Unsupervised Learning: Models uncover hidden patterns in unlabeled data, such as clustering customer preferences.
- Reinforcement Learning: Systems learn through trial and error, optimizing behavior based on rewards (e.g., AlphaGo mastering the game of Go).

Neural Networks

Inspired by the structure of the human brain, neural networks consist of interconnected layers of nodes (neurons) that process information. Key advancements include:

- **Convolutional Neural Networks (CNNs)** for image processing.
- **Recurrent Neural Networks (RNNs)** for sequential data like speech and time series.
- **Transformer Models:** Recent innovations like **GPT** and **BERT** have redefined natural language processing.

Data Science

- AI's modern success is fueled by the abundance of big data and advancements in data processing.
- Data Science integrates algorithms, statistical methods, and computational techniques to extract actionable insights.
- Cloud computing platforms and distributed systems, such as Hadoop and Spark, have made it possible to process vast datasets at unprecedented speeds.

Challenges in the Early Days

The journey of AI was not without setbacks. In its early decades, researchers faced limitations in:

- Computational Power: Early computers lacked the speed and memory for advanced AI tasks.
- Algorithmic Complexity: Foundational algorithms were primitive and struggled with real-world problems.
- Data Availability: The lack of large datasets hindered machine learning and natural language processing progress.

Despite these obstacles, the persistence of researchers and the rapid evolution of technology propelled AI into its current prominence, setting the stage for its future potential.

Looking Ahead

Understanding AI's genesis is critical to appreciating its present and future. By examining its historical context, we trace its technological roots and recognize the vision and determination of those who dared to imagine machines that could think. The story of AI is still unfolding, and its ultimate impact will depend on how we navigate its ethical, societal, and technological challenges in the years to come.

AI Today

AI Today in our Daily life

Imagine waking up to the soothing sound of a smart alarm that adjusts based on your sleep patterns, synced with your fitness tracker. As you prepare for the day, your virtual assistant provides a quick weather update on your daily schedule and even recommends the fastest route to work, avoiding traffic. All of this happens seamlessly, almost invisibly. This is the world we live in today, where Artificial Intelligence (AI) has become an integral part of our lives, transforming how we live and how industries operate.

AI in Daily Life: From Virtual Assistants to Smart Devices

AI's integration into our daily routines is so seamless that we often take it for granted. Virtual assistants like **Siri, Alexa, and Google Assistant** have become household staples, performing tasks as mundane as setting reminders to complex activities like controlling smart home ecosystems. With a simple voice command, these AI-powered systems can adjust the thermostat, lock the doors, or even order groceries, offering convenience at an unprecedented level.

Beyond virtual assistants, AI is the invisible force behind many daily services. When Netflix suggests a show you'll probably binge-watch or Amazon recommends products tailored to your preferences, it's AI at work. Streaming platforms leverage sophisticated recommendation algorithms to analyze your viewing habits and predict what you'll enjoy next. Similarly, navigation apps like Google Maps use AI to provide real-time traffic updates, optimizing your route to save time and fuel.

Smart devices, from wearables like fitness trackers to Internet of Things (IoT) gadgets, further highlight AI's pervasive influence. These devices gather data from our daily lives, process it, and deliver actionable insights, such as fitness goals or energy-saving tips for your home.

Imagine This:
A single AI ecosystem connects your smart speaker, wearable device, and home appliances. When your wearable detects that you're awake, it prompts your smart coffee maker to start brewing your favorite roast while the lights gradually brighten to ease you into the day. This is not just convenience; it's personalization driven by AI.

AI Applications in Industry

While AI enhances individual lives, its impact on industries is revolutionary. From healthcare to manufacturing, AI redefines processes, improves efficiency, and drives innovation.

Healthcare
AI has emerged as a game-changer in healthcare, offering solutions that were once considered impossible. For instance, AI-powered diagnostic tools can analyze medical images like X-rays and MRIs with astonishing accuracy, often detecting conditions earlier than human doctors. These systems aren't just fast; they're lifesaving.

AI is also reshaping patient care through personalized medicine. Imagine an AI system analyzing your genetic makeup, lifestyle, and medical history to recommend a treatment plan uniquely tailored to you. Virtual health assistants, like chatbots, are another boon. They offer round-the-clock medical advice and medication reminders, making healthcare more accessible.

Example:

An AI system in radiology processes thousands of X-rays in a fraction of the time it would take a team of doctors to identify abnormalities with precision. This enables faster diagnosis and treatment, ultimately saving lives.

Finance

AI is a vigilant watchdog in the financial sector, analyzing transaction patterns to detect fraudulent activities. For example, if your credit card is suddenly used in an unusual location or for an atypical purchase, AI algorithms often flag the transaction before you realize something's amiss.

Investment platforms, too, have embraced AI. They use machine learning models to analyze market trends, predict stock performance, and optimize portfolio management. These tools make investing more innovative and accessible, even for novice investors.

Manufacturing

Thanks to AI-driven robotics and automation, factories of the future are already here. Predictive maintenance systems monitor machinery in real time, forecasting potential breakdowns before they happen. This minimizes downtime, saving companies significant costs. On assembly lines, robots equipped with AI optimize production, improving precision and efficiency.

Ethical Dilemmas: Bias, Privacy, and Decision-Making Accountability

As much as AI offers unprecedented opportunities, it also introduces ethical challenges that demand immediate attention.

Bias

AI systems are only as unbiased as the data they're trained on. Unfortunately, data biases can result in discriminatory outcomes. For instance, AI-powered hiring algorithms have been found to favor specific demographics over others, perpetuating inequalities instead of resolving them. These issues raise fundamental questions about fairness and inclusivity in AI development.

Case Study:

A hiring AI trained on past employee data was found to favor male candidates for technical roles because historical hiring practices were skewed toward men. This unintentional bias highlighted the critical need for diverse, unbiased training datasets.

Privacy

AI thrives on data, but this dependence often leads to privacy concerns. When your smartphone tracks your location, your smart home logs your habits, or apps analyze your preferences, where does this data go? And who has access to it? The potential for misuse, from targeted advertising to surveillance, poses significant risks to individual privacy.

Accountability

When an AI-driven system makes a decision—approving a loan, diagnosing a disease, or controlling an autonomous vehicle—who is held accountable if something goes wrong? The complexity of AI systems makes it difficult to pinpoint responsibility, especially in high-stakes scenarios.

Example:

An autonomous vehicle involved in an accident raises thorny questions: Was it a failure of the AI system, the developers, or the data it was trained on? These issues highlight the urgent need for clear accountability frameworks.

AI Today: A Double-Edged Sword

Artificial Intelligence is both a marvel and a challenge today. It simplifies our lives, drives industries forward, and forces us to confront difficult ethical questions. As we navigate this new era, the decisions we make about how AI is developed, deployed, and regulated will shape the kind of future we build.

AI has made our present more innovative and efficient, but its true potential lies in the promise of tomorrow, which we must handle with care and responsibility. With each innovation, we take a step closer to a world where AI doesn't just serve us but serves the greater good.

CHAPTER THREE

The Promise of Tomorrow

The Promise of Tomorrow

The possibilities seem boundless as we stand at the threshold of an AI-driven future. Emerging technologies are pushing the limits of what AI can achieve, heralding a new era where machines might assist and collaborate with us in solving humanity's most significant challenges. From Artificial General Intelligence (AGI) to quantum computing and neural-symbolic AI, advancements promise to revolutionize industries, address global crises, and unlock untapped creative potential. This chapter explores the technologies, opportunities, and profound impact AI holds for tomorrow.

Emerging AI Technologies

Artificial General Intelligence (AGI): A Leap Towards Human-Like Intelligence

The concept of AGI represents the ultimate aspiration of AI researchers—a machine with cognitive abilities that rival human intelligence. Unlike narrow AI, which excels at specific tasks (e.g., playing chess or analyzing images), AGI aims to emulate human versatility, enabling machines to learn, reason, and make decisions across various domains.

AGI is still a theoretical pursuit, with significant obstacles to overcome, such as developing algorithms capable of generalizing knowledge and achieving self-awareness. Yet, progress in fields like deep reinforcement learning, transfer learning, and natural language processing signals that AGI is not just science fiction but a future possibility.

Imagine This:

An AGI system that learns a new language as effortlessly as a child designs a sustainable city and collaborates with scientists to solve the mysteries of dark matter—all in a single day. While we are far from this reality, each breakthrough brings us closer to machines capable of reshaping the human experience.

Quantum Computing: The Catalyst for Exponential Growth

Quantum computing is poised to revolutionize AI by addressing problems too complex for classical computers. Unlike traditional bits, which exist in binary states (0 or 1), quantum bits, or qubits, can exist in multiple states simultaneously, enabling vast parallel processing.

This capability holds transformative potential for AI, particularly in areas requiring massive computational power, such as:

Cryptography: Developing unbreakable encryption methods.

Optimization: Solving logistical challenges in supply chains and transportation.

Simulations: Modeling biological systems for drug discovery or climate predictions.

While quantum computing is still in its infancy, companies like IBM, Google, and D-Wave are rapidly advancing the technology. This is paving the way for AI systems capable of tackling previously unsolvable problems.

Neural-Symbolic AI: Merging Learning with Reasoning

Like neural networks, traditional AI systems excel at pattern recognition but often struggle with reasoning and logic. On the other hand, symbolic AI focuses on rule-based reasoning but cannot learn from data. Neural-symbolic AI bridges this gap by combining the strengths of both approaches.

This hybrid model enhances AI's ability to generalize knowledge, interpret abstract concepts, and apply logical reasoning. It is particularly valuable in fields like law, where reasoning and contextual understanding are paramount. Neural-symbolic AI represents a significant step toward making AI more explainable and reliable.

Example:
A neural-symbolic AI system could analyze medical literature (symbolic reasoning) and learn from patient data (neural networks) to provide precise diagnoses while justifying its recommendations with evidence.

The Potential for Solving Global Challenges

Climate Change
AI offers powerful tools to combat one of humanity's most pressing crises—climate change. AI-powered predictive models can guide policies and actions to mitigate environmental damage by analyzing complex climate patterns. For example:

AI-driven energy optimization systems improve the efficiency of renewable energy sources like wind and solar.

Smart grids, powered by AI, balance energy supply and demand, reducing waste and emissions.

AI models predict the impact of rising sea levels, helping communities prepare for environmental changes.

Healthcare

The healthcare sector is transforming with AI as its catalyst. From diagnostics to drug discovery, AI is improving outcomes, reducing costs, and expanding accessibility:

Accelerating Drug Discovery: AI algorithms analyze molecular structures, identifying potential drug candidates in weeks rather than years.

Telemedicine: AI-powered virtual health assistants offer medical consultations and monitor patient conditions, particularly in remote areas with limited healthcare access.

Case Study:

During the COVID-19 pandemic, AI systems accelerated the development of vaccines by analyzing vast amounts of data on the virus's genetic structure, exemplifying how AI can respond to global emergencies.

Education

AI is democratizing education by making it more personalized and accessible. Adaptive learning platforms assess individual student needs and tailor lessons to optimize learning outcomes. Language translation tools and AI-powered tutors break down barriers, ensuring that knowledge reaches every corner of the world.

Imagine an AI system that teaches and learns from students, adapting its methods to suit different learning styles and fostering an environment where every student can thrive.

The Role of AI in Augmenting Human Creativity and Innovation

AI's capabilities extend far beyond automation; it is becoming a collaborator in human creativity. By generating ideas, enhancing workflows, and expanding possibilities, AI inspires new forms of artistic and scientific innovation:

Art

Tools like Generative Adversarial Networks (GANs) enable AI to create stunning visual art that challenges traditional notions of creativity. These systems don't just mimic existing styles—they generate entirely new ones. Artists now collaborate with AI to push the boundaries of their craft.

Music

AI-powered composition tools allow musicians to explore novel soundscapes. These tools can create personalized compositions tailored to specific moods or occasions by analyzing vast musical datasets.

Science

In scientific research, AI accelerates breakthroughs by processing complex datasets, identifying patterns, and generating hypotheses. For instance, AI systems are instrumental in modeling protein structures, a critical step in developing disease treatments.

Example:

The AI-generated artwork "Edmond de Belamy," sold at auction for over $400,000, sparked debates about the role of AI in creativity and its potential to reshape industries.

The Promise of Tomorrow

The promise of AI is not just about building more intelligent machines—it's about creating a more innovative world. Whether it's addressing global crises, augmenting human creativity, or pushing the boundaries of innovation, AI can unlock a more efficient, equitable, and imaginative future.

Yet, with great power comes great responsibility. As we explore the vast potential of emerging AI technologies, we must ensure they are developed and deployed ethically, inclusively, and sustainably. The promise of tomorrow lies not in the technology itself but in how we choose to use it.

A Call to Action:

The technologies shaping tomorrow's AI are within our reach. As you read this chapter, consider the opportunities and challenges that come with them. The future is being written today—and AI is poised to be one of its greatest authors.

CHAPTER FOUR

Challenges and Risks

Challenges and Risks

As Artificial Intelligence continues to advance and integrate into nearly every facet of society, it brings with it not only transformative possibilities but also profound challenges and risks. The power of AI lies in its ability to augment human capabilities and solve complex problems, but unchecked, it can also amplify biases, erode privacy, and exacerbate societal inequalities. This chapter delves into the ethical and societal challenges, the risks posed by autonomous systems, and the urgent need for robust governance and regulation to ensure AI serves as a force for good.

Ethical and Societal Challenges

Algorithmic Bias: A Reflection of Society's Prejudices

One of AI's most pressing ethical concerns is its potential to reinforce or even magnify societal biases. AI systems learn from historical data, and when this data contains prejudices—whether in hiring practices, lending decisions, or policing—AI models can perpetuate or amplify these inequities. For example, facial recognition systems have been shown to perform poorly on darker skin tones, leading to false identifications and discriminatory outcomes.

Bias in AI is not just a technical flaw; it is a societal issue that affects real lives. A hiring algorithm, for instance, trained on past hiring data, may favor specific demographics over others, inadvertently excluding qualified candidates based on gender, race, or socioeconomic background. These outcomes challenge our values of fairness and equality and demand immediate attention.

Impact:

- Loss of trust in AI systems.
- Legal and ethical ramifications for organizations deploying biased AI.
- Deepening societal divisions.

Privacy Concerns: The Cost of Data-Driven Intelligence

AI thrives on personal, behavioral, and contextual data. However, this data's collection, storage, and use often occur without explicit consent, raising significant privacy concerns.

Imagine an AI-powered smart home device listening to conversations or tracking your daily routines. While these systems enhance convenience, they also pose risks of surveillance, data breaches, and third-party misuse. The erosion of privacy undermines individual freedoms, creating a society where people are constantly monitored, intentionally or unintentionally.

Key Examples:

- Location tracking by apps without user consent.
- AI models predict personal attributes (e.g., sexual orientation or political affiliation) from seemingly innocuous data like browsing habits or social media activity.

Impact:

- Loss of autonomy over personal information.

- Vulnerability to identity theft and cyberattacks.
- Widespread mistrust of AI technologies.

Job Displacement and Economic Inequality

AI-driven automation is reshaping the workforce, replacing repetitive and manual tasks with machines capable of working faster, cheaper, and more accurately. While this shift boosts productivity, it displaces human labor, particularly in manufacturing, retail, and logistics.

The rise of AI poses significant challenges for workers whose jobs are vulnerable to automation, particularly those with low-skill or routine-based roles. The resulting economic inequality could widen the gap between those who benefit from AI (e.g., tech-savvy professionals) and those who are left behind.

Consider This:

An autonomous delivery fleet will replace human drivers, eliminating thousands of jobs while generating profits for a few tech companies.

Impact:

- Increased unemployment in vulnerable sectors.
- Economic disparities between skilled and unskilled workers.
- There is a growing need for reskilling and education initiatives.

The Risks of Autonomous Systems

The Double-Edged Sword of Autonomy

AI-powered systems, from self-driving cars to autonomous drones, promise unparalleled efficiency and innovation. However, autonomy also introduces unique risks, particularly when these systems fail or behave unpredictably.

For example, self-driving cars rely on AI to make split-second decisions, but determining accountability becomes complex when these decisions result in accidents. Was it a flaw in the AI's design, the manufacturer's oversight, or the operator's misuse? Such incidents not only lead to loss of life and property but also raise fundamental questions about trust and responsibility.

Case Study:

In 2018, an autonomous vehicle in testing struck and killed a pedestrian. The incident highlighted flaws in the car's object detection system and raised legal and ethical questions about accountability.

Risks Include:

- Accidents: Failures in decision-making can lead to catastrophic consequences.
- Security Vulnerabilities: Autonomous systems can be hacked, manipulated, or weaponized.
- Moral Dilemmas: AI in autonomous systems may face ethical quandaries, such as deciding who to prioritize in a life-threatening scenario.

Regulation and Governance of AI

The Need for a Global Framework

As AI becomes more powerful and pervasive, the absence of robust governance frameworks poses significant risks. International collaboration is essential to establish safety, ethics, and transparency standards while fostering innovation. Without regulation, AI could exacerbate societal inequalities, violate human rights, or even be weaponized in conflict zones.

Key Elements of AI Governance:

1. **Ethical Guidelines:**

- AI systems must prioritize fairness, inclusivity, and accountability.
- Developers should ensure that AI is free from harmful biases and respect cultural diversity.

2. **Transparency and Explainability:**

- Algorithms must be transparent, with clear documentation of how decisions are made.
- Users should understand why an AI system made a specific recommendation or decision.

3. Accountability Mechanisms:

- Clear protocols to address harm caused by AI, including compensation and legal responsibility.
- Shared accountability among developers, manufacturers, and operators.

4. Regulatory Oversight:

- Independent regulatory bodies will monitor and evaluate AI technologies.
- Periodic audits of AI systems to ensure compliance with ethical and safety standards.

Proposed Governance Framework

A successful governance framework must strike a delicate balance between fostering innovation and ensuring public safety.

Key Components:

- International Collaboration: Cross-border initiatives to align standards and share knowledge.
- AI Ethics Committees: Multidisciplinary teams will evaluate the ethical implications of AI projects.
- Public Awareness Campaigns: Educating society on the benefits and risks of AI to foster informed decision-making.

The Crossroads of Promise and Peril

AI is a powerful tool, but with great power comes great responsibility. This chapter outlines the challenges and risks, highlighting the urgent need for thoughtful, inclusive, and proactive AI development and deployment approaches.

As we navigate these challenges, one thing is clear: the future of AI will not be determined solely by technological advancements but by the ethical and societal choices we make today. Only by addressing bias, safeguarding privacy, and establishing robust governance can we ensure that AI fulfills its promise as a force for good in the world.

CHAPTER FIVE

A Vision for the Future

A Vision for the Future

The emergence of Artificial Intelligence has prompted profound questions about humanity's future. Will AI dominate our lives, or will we harness it as a powerful tool for collaboration, creativity, and progress? To ensure a harmonious coexistence with AI, we must proactively shape its role in our society, aligning it with our values, aspirations, and collective well-being. This chapter explores how humanity can coexist with AI, how work and leisure may be redefined, and what steps are necessary to prepare future generations for an AI-centric world.

How Humanity Can Coexist with AI

Collaborative AI: Partners, Not Replacements

The key to successful coexistence is designing AI systems that complement human capabilities rather than replacing them. Collaborative AI envisions humans and machines working together, each contributing their unique strengths. AI excels at processing vast amounts of data and performing repetitive tasks, while humans bring creativity, empathy, and ethical judgment.

In fields like healthcare, collaborative AI can enhance outcomes by assisting doctors in diagnostics while leaving critical decision-making in human hands. Similarly, AI can generate innovative designs in industries like architecture while architects refine them to meet human needs and aesthetics.

Consider This:

Imagine a future where an AI-powered assistant helps a teacher design customized lesson plans for each student. This partnership would free the teacher to focus on inspiring and mentoring the class, amplifying human potential without eroding human agency.

Building Trust Through Education and Transparency

Public trust in AI is crucial for its acceptance and integration into society. Misunderstandings about AI's capabilities and concerns over misuse often breed fear and resistance. Addressing this requires:

- Transparency: Clear explanations of how AI systems work and their decisions.
- Engagement: Open dialogues between developers, policymakers, and the public to demystify AI and address concerns.
- Education: Empowering individuals with knowledge about AI's role and limitations, fostering a sense of shared responsibility.

By fostering trust and understanding, we can create an environment where AI is viewed as a partner in progress rather than a threat to humanity.

Redefining Work and Leisure in an AI-Driven World

The Future of Work: A Shift to Creativity and Strategy

As AI takes over routine and repetitive tasks, work will transform. Jobs that involve data entry, fundamental analysis, or manual assembly are increasingly automated. Still, this shift allows humans to focus on roles requiring creativity, strategic thinking, and emotional intelligence.

To navigate this transition, society must reimagine job structures and prioritize skills machines cannot replicate. Fields like arts, innovation, caregiving, and leadership will become more valuable. This transition also raises the question of economic stability for displaced workers, prompting discussions about alternative models like Universal Basic Income (UBI) or government-supported retraining programs.

Imagine This:

An AI-powered assistant manages repetitive administrative tasks for a marketing team, allowing team members to focus on crafting compelling campaigns and building meaningful client relationships.

The Evolution of Leisure

Immersive technologies and personalized experiences will enhance leisure in an AI-driven world. AI-driven platforms can curate activities, hobbies, and entertainment tailored to individual preferences. For instance, AI-powered virtual reality (VR) environments will create lifelike, interactive storytelling experiences that blend entertainment and education.

Examples of AI-Enhanced Leisure:

- Personalized playlists and movie recommendations based on mood and past behavior.
- AI-guided virtual tours of historical sites offering rich, interactive narratives.
- Hobby assistants, such as AI art tutors or music composition tools, help individuals unlock creative potential.

In this future, leisure becomes more enriching and accessible, catering to diverse interests and preferences.

Preparing Future Generations for an AI-Centric Society

Rethinking Education for the AI Era

Education systems must evolve to emphasize skills that machines cannot replicate to thrive in a world increasingly shaped by AI. This includes:

- Digital Literacy: Understanding how AI works and how to interact with it responsibly.

- Critical Thinking: Developing the ability to analyze, question, and make informed decisions.
- Emotional Intelligence: Strengthening interpersonal skills, empathy, and leadership abilities.

Encouraging interdisciplinary learning, where technology intersects with the humanities, will help students appreciate AI's potential and ethical considerations. For example, combining computer science with philosophy can nurture developers to prioritize fairness and inclusivity in their designs.

Case Study:

A school in Finland integrates AI into its curriculum, teaching students to code while discussing the ethical implications of AI in society. This approach produces graduates who are both technically skilled and socially conscious.

Lifelong Learning and Workforce Adaptation

AI's rapid evolution necessitates continuous learning throughout life. Governments, institutions, and organizations must invest in programs that reskill and upskill workers for emerging industries. Lifelong learning opportunities, such as online courses and vocational training, will ensure that no one is left behind in the AI revolution.

The Ethical Foundations for the Future

As we prepare for a future in which AI plays a central role, it is imperative to establish strong ethical foundations. Future generations must be taught to prioritize values like fairness, accountability, and sustainability in AI development and deployment. This requires a collective effort from educators, policymakers, and industry leaders to ensure that AI is a tool for equitable progress.

A Society Thriving on Innovation and Collaboration

The vision of the future is not one of humans versus machines but of humans and machines together, creating a society where AI complements human potential. By embracing collaboration, redefining work and leisure, and preparing future generations, we can shape a world where technology enhances rather than diminishes our humanity.

The promise of tomorrow lies in our ability to adapt and evolve, leveraging AI as a catalyst for innovation, equity, and shared progress. With thoughtful planning and collective effort, we can build a future where AI empowers us to reach new heights while staying true to our core values.

• 31 •

Epilogue

As we close the final chapter of Tomorrow AI, it is clear that artificial intelligence is not just a technological marvel but a defining force shaping our collective future. From the intricate algorithms that mimic human thought to the profound transformations across industries, AI is no longer confined to the realm of science fiction. It is here, it is now, and it is ours to guide. Writing this book has been both a journey and a revelation. It has reaffirmed my belief that the interplay of human ingenuity and machine intelligence holds the key to solving some of the world's most pressing challenges. Yet, it has also underscored the responsibility we bear—as individuals, as communities, and as a global society—to ensure this technology serves the greater good. The future is not something that happens to us; it is something we create. As you close this book, I invite you to carry forward the questions, ideas, and possibilities we've explored. How will you contribute to a future where AI amplifies human potential? How will you ensure that ethics, inclusivity, and sustainability remain at the forefront of this unprecedented evolution? I hope this book has inspired you to look at the world with a renewed sense of curiosity, optimism, and responsibility. Together, we have the power to shape a tomorrow where artificial intelligence uplifts humanity, strengthens our values, and unlocks opportunities we have yet to imagine.

The story of AI is just beginning, and we are all its co-authors. Let's write a future we can be proud of. Thank you for being a part of this journey.

With hope and determination,
Ramesh Pingili

Afterword

As we journey through the age of Artificial Intelligence, one truth becomes abundantly clear: AI is not just a technology but a reflection of who we are and what we aspire to become. It holds a mirror to our values, decisions, and priorities, reminding us that its power lies not in its algorithms but in the hands of those who create and wield it.

Throughout this book, we've explored AI's genesis, its transformative presence in our lives today, and the boundless possibilities it offers for tomorrow. We've also confronted its challenges—ethical dilemmas, societal implications, and the risks inherent in its rapid advancement. These dualities—promise and peril, opportunity and responsibility—define AI's role in our collective future.

But this story is far from over. The evolution of AI is a shared journey that requires collaboration between technologists, policymakers, educators, and society at large. Together, we must navigate the complex interplay between innovation and ethics, progress and equity, ensuring that AI is a tool for empowerment rather than division.

This book is not the conclusion of a narrative but the beginning of a dialogue—a call to action for individuals, industries, and nations to engage thoughtfully with the challenges and opportunities ahead. Whether you are a developer pushing the boundaries of what AI can achieve, a policymaker shaping its governance, or simply a curious mind pondering its impact, your role is vital in shaping the future.

As we close these pages, I invite you to imagine the world we can create together—where AI amplifies human potential, solves humanity's most significant challenges, and inspires new forms of creativity and connection. That world is not a distant possibility; it is a vision we can achieve, starting today.

Thank you for joining me on this exploration of Tomorrow AI. Let us step into the future with curiosity, courage, and a shared commitment to making it brighter for all.

With hope and gratitude,

Ramesh Pingili

IEEE Senior Member, Fellow of IETE, Full Member of Sigma Xi

Technologist and Author Specializing in AI and Workflow Optimization

AFTERWORD

Glossary Of Terms

This glossary defines key terms and concepts related to artificial intelligence (AI), which this book discusses. It is designed to help readers understand the technical aspects of AI and its applications.

Algorithm

A step-by-step procedure or set of rules designed to perform a specific task or solve a problem. In AI, algorithms form how machines process information and make decisions.

Artificial General Intelligence (AGI)

AGI is a type of AI that aspires to achieve human-like intelligence. It enables machines to learn, reason, and perform any intellectual task that a human can do. However, AGI is still a theoretical concept.

Artificial Intelligence (AI)

The simulation of human intelligence in machines programmed to think, learn, and make decisions. AI systems can perform problem-solving, natural language understanding, and pattern recognition.

Autonomous Systems

AI-powered systems that can operate independently without direct human intervention. Examples include self-driving cars, drones, and robotic process automation.

Bias in AI

AI systems tend to produce results that favor or discriminate against certain groups. Bias often occurs due to imbalances or prejudices in the data used to train the AI model.

Collaborative AI

AI is designed to work alongside humans, complementing their abilities rather than replacing them. Collaborative AI enhances productivity while maintaining human oversight.

Data Science

An interdisciplinary field that uses algorithms, statistical models, and computational tools to analyze and interpret large datasets. It plays a crucial role in training AI systems.

Deep Learning

Deep learning is a subset of machine learning that uses multi-layered neural networks to process data and learn from patterns. It powers advanced applications like image recognition and natural language processing.

Generative Adversarial Networks (GANs)

A type of AI system where two neural networks work against each other: one generates data while the other evaluates its authenticity. GANs are often used to create realistic images and videos.

Machine Learning (ML)

ML is a subset of AI that enables machines to learn and improve from experience without being explicitly programmed. ML techniques include supervised learning, unsupervised learning, and reinforcement learning.

Neural Networks

Neural networks are a type of AI architecture inspired by the structure of the human brain. They consist of interconnected nodes (neurons). Neural networks process information in layers to recognize patterns and make decisions.

Neural-Symbolic AI

A hybrid approach that combines the learning capabilities of neural networks with the logical reasoning of symbolic AI. This method enhances the system's ability to understand abstract concepts.

Predictive Modeling

AI analyzes statistical data and makes predictions about future outcomes. It is widely used in healthcare, finance, and climate science fields.

Quantum Computing

Quantum computing is a computing paradigm that uses quantum-mechanical phenomena, such as superposition and entanglement, to perform calculations at speeds far beyond those of classical computers. It has the potential to revolutionize AI.

Reinforcement Learning

A type of machine learning where an AI agent learns by interacting with its environment and receiving feedback through rewards or penalties.

Supervised Learning

A machine learning technique where models are trained on labeled data, meaning the input and corresponding correct output are provided. The system learns to predict the output for new inputs.

Turing Test

A test proposed by Alan Turing to determine if a machine can exhibit intelligent behavior indistinguishable from that of a human. Passing the Turing Test is considered a milestone in AI development.

Universal Basic Income (UBI)

An economic model proposing regular, unconditional payments to individuals intended to address income disparities and job displacement caused by automation and AI.

Virtual Assistants

AI-powered applications like Siri, Alexa, and Google Assistant perform tasks and provide information through voice or text-based interactions.

This glossary serves as a quick reference for readers to understand the technical terminology used throughout the book and deepen their engagement with the concepts of AI.

GLOSSARY OF TERMS

About The Author

Ramesh Pingili

Ramesh Pingili is a renowned technologist and thought leader with deep expertise in artificial intelligence, workflow optimization, and robotic process automation. An IEEE Senior Member and a Fellow of the Institution of Electronics and Telecommunication Engineers (IETE), Ramesh Pingili has played a pivotal role in advancing enterprise technology solutions that drive efficiency and innovation. The author is also the creator of www.pegahelp.com, a leading technology blog that offers invaluable insights and guidance to professionals and organizations adopting cutting-edge Pega systems. Their dedication to knowledge sharing extends to key research publications in esteemed journals like IJRCAIT, IJCET, and IJSRCSEIT, highlighting transformative contributions to AI and related fields. Through https://rameshpingili.com/, Ramesh Pingili shares personal reflections, professional milestones, and innovative perspectives on technology's future. Their work focuses on ethical AI and advocates for the harmonious integration of technology and society to foster sustainable progress. "Tomorrow AI" embodies their vision of an intelligent, connected future and is a testament to their unwavering commitment to education, innovation, and thought leadership

A Final Thought

As we embrace the age of Artificial Intelligence, we are reminded that the future is not something that happens to us—it is something we actively create. AI, in essence, reflects our ingenuity, values, and collective ambition to push the boundaries of what is possible. It is a tool that holds the power to solve our most significant challenges, amplify our creativity, and bring us closer as a global society.

Yet, this power comes with a profound responsibility. The choices we make today—how we develop, deploy, and govern AI—will shape the legacy we leave for future generations. Let us strive to ensure that AI is guided by the principles of equity, empathy, and ethical integrity, fostering a world where technology serves humanity and not the other way around.

The journey of Tomorrow AI is just the beginning of a larger conversation—a call to think critically, act responsibly, and dream boldly. Together, we can build a future where AI and humanity coexist harmoniously, unlocking potential we have only just begun to imagine.

The future is bright. Let's step into it with hope, curiosity, and a shared vision for a better tomorrow.

With optimism for what lies ahead,
Ramesh Pingili